Two Sussex archaeologists- William Durrant Cooper and Mark Antony Lower

HENRY CAMPKIN

1877

TABLE OF CONTENTS

WILLIAM DURRANT COOPER

The year 1812, in the very dawn of which the subject of this imperfect sketch first saw the light, was one of the most eventful, most memorable years of the nineteenth century. In that year, as is well known, "the scourge of Europe," the first Napoleon, was at last effectually checked in his career of conquest and confiscation. In England the high price of provisions and scarcity of work, and the distress and discontent consequent thereon, led to continuous local disturbances and riotings, and the wholesale destruction of machinery. Unhappy rioters, or so-called rioters, were hanged, half-a-dozen or more at a time. On one occasion, eight poor ignorant wretches were thus disposed of at Manchester, one of them being a miserable woman, whose sole offence was the stealing of a few potatoes. In 1812, too, a cabinet minister Spencer Perceval was assassinated in the lobby of the House of Commons. And if to this it be added that the United States of America declared war against England, and in several instances compelled English ships, after hard fights, to strike their flags to their transatlantic assailants, it will be seen that, taking it altogether, the year 1812 was as gloomy and unpromising a one as a human being could well choose or have chosen for him for his entry upon the theatre of life.

Mr. Cooper's ancestry may be traced back to Thomas Cooper, of Icklesham, a Sussex squire of the seventeenth century. Thomas Cooper, his eldest great-grandson, also of Icklesham, who married, in 1787, Mary, daughter of Thomas Collins, of Winchelsea, had six sons and two daughters. The second of these six sons was Thomas Cooper, who, born in May 1789, married Lucy Elizabeth, great-granddaughter of Samuel Durrant, of Cockshot, Hawkhurst, Kent; and the eldest son of this marriage was

William Durrant Cooper, who was born in High Street, in the parish of St. Michael, Lewes, on the tenth of January, 1812. The first cadet of this family, who settled in Lewes, would seem to have been William Cooper, the second of the great-grandsons of the first-named Thomas Cooper, of Icklesham. He became an eminent solicitor in Lewes, and dying in 1813, was succeeded in his practice by his nephew, Thomas Cooper, the father, as just stated, of the subject of this notice. This William Cooper was perhaps the only member of the legal profession who espoused the Liberal side of politics in Lewes. His residence was in Saint Anne's parish, and being well-nigh as independent in pocket as he was in politics, and endowed, moreover, with a spice of humour as well, he could afford to indulge in a practical joke upon his electioneering opponents, without counting its cost too nicely. In connection with Sir Henry Blackman, these two being the chief supporters of what was called the independent party in Lewes, he brought forward Mr., afterwards Sir James, Scarlett, and subsequently Lord Abinger, on the first occasion that eminent lawyer, then a flaming Whig, and afterwards a more flaming Tory, contested, unsuccessfully, the old parliamentary borough, which then had the privilege of returning two members. With no greater success Mr. Scarlett ventured on a second contest. On the first of these contests (1812) he lost his election by nine votes. On the second (1816) he was in a minority of nineteen.

After the 1812 contest Mr. William Cooper, incensed at the conduct of all the butchers of the town, who, like all the lawyers of the town, except himself, voted against his chosen candidate, hit upon the novel vengeance of opening an opposition butcher's shop in Saint Anne's, painting over it, in conspicuous letters, "William Cooper, Butcher," and under-sold the blue-aproned trade in their own commodities, at the rate of one penny per pound a consideration in those dear days until they capitulated, and, as the story goes, promised to support his candidate at the next election; a pact which, if entered into, can hardly have been adhered to, as we see above that Mr. Scarlett found at that next election the majority against him had increased from nine to nineteen. Possibly, Mr. William Cooper having died in 1813, the butchers aforesaid deemed themselves released by his death from the performance of their forced promise.

William Durrant Cooper took his first Christian name, from his great-uncle, the just-mentioned practical joker, who was his godfather; his second name, being, as already stated, his mother's maiden name. He received his education at the Grammar School, Lewes, whose head-master, for all the latter time of his stay there, was Dr. George Proctor, afterwards principal of Saint Elizabeth's College, Guernsey, and now the venerable Chaplain to the Fishmongers' Almshouses at Bray, near Maidenhead, Berks. While subject to Dr. Proctor's direction, this local Grammar School attained a high character, and under him, Mr. Cooper, for whom his tutor always entertained a high regard, early showed great intelligence, and made rapid progress in his studies. But from this ancient seminary, his only alma mater, he was perhaps too prematurely taken, for he was not more than fifteen years of age when he was articled as clerk to his father, and during his articles, although he may not literally have realised Pope's couplet, and have been

"A clerk foredoomed his father's soul to cross,
Who pens a stanza, when he should engross,"

he yet exhibited an early bias towards literature, but the severer Clio modern scholiasts write the name Cleio rather than those of her sisters who dallied with poetry in its various forms, was the Muse to whom his youthful heart was vowed, and unto whom, through life, his multifarious labours were chiefly dedicated. History history in its topographical and archaeological phases was the study in which he delighted, and he was not out of his teens ere the history and antiquities of his native town and county engaged his constant and serious attention, and as time rolled on, he made himself familiar with those of most of the Sussex families of any local importance. He not only materially assisted Mr. Horsfield in the compilation of his History of Sussex, but, while he was not an author on his own account, at so early an age as his friend Lower, still, by the time he had completed his twenty-second year, that is in 1834, he had contributed a valuable supplement to Mr. Horsfield's work, under the title of The Parliamentary History of the County of Sussex, and of the several Boroughs and Cinque Ports therein. This Parliamentary history of his native County, which was also issued in a separate form, compressed into fifty-three double-column quarto pages of very small type, would readily fill a respectable octavo volume, and, as to the way in which it is executed, would reflect credit, both for its painstaking and research, upon the most experienced historian.

In 1836 Mr. Cooper published A Glossary of the Provincialisms in use in the County of Sussex. This slim volume, which was "printed for private distribution" only, and probably first appeared in the columns of the Brighton Herald, from the office of which it emanated in its book shape, has since been thrown into the shade by the more comprehensive Glossary, issued a year or two ago, by the Rev. W. D. Parish, the learned Vicar of Selmeston, who, as a diligent labourer in the same field, would certainly be among the first to appreciate the efforts of his predecessor.

In 1842 he published The Sussex Poets, a lecture at Hastings. This little brochure has, in all probability, been long out of print. In the strict order of events, it ought to have been sooner stated, that—if the Law List be correct—previous to the completion of his twenty-first year, namely, in Michaelmas Term, 1832, he was duly admitted an attorney and solicitor.

It would have been strange if, with his peculiar bias, the disgracefully neglected state of our Parish Registers, so much excitement about which prevailed some forty or fifty years ago, had not, even from a professional point of view, forced itself on Mr. Cooper's attention. Accordingly, with his usual activity, he bestirred himself in the matter, and in April, 1833, when he had turned his twenty-first year by three months only, he was called before the House of Commons Committee, then sitting, on Parochial Registration, to give his young, but by no means immature, experience on the condition, mostly, of the registers of his own County; and the state of things disclosed in his evidence, which covers eight printed folio pages, reflected great discredit on the previous contemporary custodians of those precious records. He had seen, in the difficulties thus interposed in the clearing up of titles on the sale or purchase of landed property, proof positive of the evils of the existing system, or rather no-system, and he exposed them most unsparingly. In his evidence as to the reckless carelessness with which the registers were treated, he states that he recollected " an instance where the clerk was about destroying the old register, saying it was of no use;" and he recollected also, "when a little boy, the parish clerk of another parish saying, that the clergyman used to direct his pheasants with the parchment of the old registers." And he was wont to relate that, once, when he went to make a search, the first sight that caught his eye, on entering the parsonage house, was a little boy riding cock-horse across a walking-cane, with a parchment cap on his head, made from a leaf of the Register. It cannot be concealed that among some of the old-school "clerics," and their deputies, Mr. Cooper's popularity was not increased by his denunciation of their disregard of the sacredness of their trust in this respect.

Emulating from the outset the conduct of his father and great-uncle, he at once heartily espoused the principles of the Liberal Party, and soon became associated with its local leaders, and no one, who was at all intimate with him, will require to be told that he became a most energetic participator in the numerous election contests of his time. He acquired so great a proficiency in election law as to be regarded as a safe authority therein, and always displaying great courage and talent, he generally won the applause of his opponents, even when they were on the losing side. Indeed, his old friend, and, practically, his first legal tutor, Mr. John Smith, then the managing clerk to Cooper père, and now the veteran actuary of the Lewes Savings' Bank, always lamented his pupil's too eager devotion to the interests of his party, as he thereby barred the way to that degree of pecuniary independence to which, with a less prominent intermingling in electioneering strife, his unquestionable talents and persevering habits

would have conducted him. But, like Milton, his inborn predilections and
too pronounced opinions would not allow him to

"...... take the beaten path and broad,
Which leads right on to fortune."

In or about the year 1837, Mr. Cooper went permanently to reside in
London, chiefly, it is believed, at the invitation of the late Sir John
Easthope, Bart. who (then plain Mr. Easthope) fought a losing battle for
the seat rendered vacant by Mr. Kemp's retirement. Sir John was the
principal, if not the sole, proprietor of the now defunct Morning Chronicle,
and Mr. Cooper, in addition to his endeavours to establish himself in his
profession, accepted a post on the parliamentary staff of that (in its day)
influential Whig journal. After a while he accepted similar employment on
the Times, but some new division of labour in the corps of reporters on the
establishment of that leviathan broad-sheet, which would have interfered
with his allotment of the daytime to his professional practice, ultimately led
to his severance from a journalistic career.

The branches of his profession in which Mr. Cooper was chiefly engaged,
were conveyancing and parliamentary agency, but it may be added that his
practice was at no time extensive, and he consequently never realised more
than a modest income.

On the death of his uncle, Mr. Frederick Cooper, who was private solicitor
to the then Duke of Norfolk, Mr. Cooper was appointed the Duke's
steward of the Leet Court of the Borough of Lewes. It is not needful to say
much here of the antiquity or jurisdiction of this "Lewes Leet," as it is curtly
styled, but, it would seem, by descent or partition among coheiresses, the
lordship of the Leet has come to be divided among the Duke of Norfolk,
the Marquis of Abergavenny, and the Earl Delawarr, the Marquis holding
two fourth parts, and the other two noble personages one fourth part each,
and the annual holding of the Leet is presided over by their stewards
alternately, the Marquis, in right of his two-fourths, being the lord for two
years in succession. A jury is summoned at each leet, and this jury presents
the names of the High Constables and Headboroughs for the ensuing year,
and, according to ancient custom, the Steward accepts the nomination thus
made, and the officers so nominated are sworn in by him. Other occasions
also arise on which leets are held, and the small fees payable to the steward
constitute the principal, if not the sole, emoluments of his office, and it may
well be imagined that the prestige attaching to the post is of quite as much
importance as its pecuniary profits. Mr. Cooper, no doubt, valued this
appointment for the periodical opportunity it afforded him of keeping up
his connection with his native town, and with his old friends there; and as

the business of the chief day was terminated by a pleasant dinner, the conviviality which then ensued, we may be sure, was not the least agreeable feature of the Lewes Leet.

Another post, not of a public character, to which Mr. Cooper succeeded in 1843, and upon which he set a high value, was the auditorship of an ancestral estate in the district of Cleaveland, in the North Riding of Yorkshire, whereon stands Skelton Castle, " a noble embattled mansion presenting a very extensive front," on whose site formerly stood "an ancient fortress, built, soon after the Conquest, by Robert de Brus, from whom descended some of the Scottish kings." Readers of the Sussex Archaeological Collections will hardly need to be reminded of the connection of the Brus family (with its various spellings of Braose, Braoze, Breuze, Brewes, Brewis, Brewose, Brewosa, Brewus, Brewys,andc.) with our southern county, and its large holdings therein. Adam de Brus, one of the early owners of Skelton Castle, on the marriage of his only daughter, Isabel, with Henry de Perci, lord of Petworth, gave to the latter a manor in Cleaveland, on the condition that "the said Henry and his heirs should repair to Skelton Castle every Christmas day, and lead the lady of that castle from her chamber to the chapel to mass, and from thence to her chamber again, and after dining with her to depart." As Skelton Castle is distant from Petworth a good three hundred miles and more, Henry de Perci and his successors must have had many a perilous and weary winter jaunt, to fulfil the condition of the tenure of this manor. This custom has, of course, long ceased, but, although centuries have passed away, Skelton Castle is still possessed by a worthy descendant of its original owner, Robert de Brus, uncle of the just named Adam de Brus. Nor is this all. Skelton Castle is a potent entity in the estimation of every admirer of Laurence Sterne, for it is the Crazy Castle of that most original (and, perhaps, most plagiaristic) of our great English authors, and his Eugenius was none other than the castle's then owner, John Hall Stevenson, himself the author of three humorous volumes a shade too free, it may be, for the present generation entitled "Crazy Tales."

Mr. Cooper had not long taken upon himself the auditorship of Skelton Castle, before his good genius instinctively led him to its muniment room, where he soon dug down upon some precious relics of " poor Yorick," which he printed, with annotations from his own critical pen, with this title page: Seven Letters, written by Sterne and his Friends, hitherto unpublished. Edited by William Durrant Cooper, F.S.A. London. Printed for private circulation, 1844. A happy Sterne-ophilist is he who possesses a copy of this rare fasciculus.

The most ambitious work, in a separate form, published by Mr. Cooper, is his History of Winchelsea, one of the Ancient Towns added to the Cinque Ports. This history appeared in 1850. Its value is testified to by the fact, that, although of so comparatively recent an issue, it is a volume rarely to be obtained. The two papers on Winchelsea, by Mr. Cooper, in Vols. viii. and xxiii. of the Sussex Archaeological Collections, form an apt complement to this volume.

On the 20th of December, 1858, Mr. Cooper was appointed to the office of Solicitor to the Vestry of Saint Pancras, Middlesex. He had previously approved himself a likely person for such an office, by the interest he had taken in, and the assistance he had given to, the passing of the Metropolitan Burials Bill, in 1852 a measure of great importance to so large and densely populated a parish as that of Saint Pancras. The emoluments of this post, consisting partly of a salary and partly of fees, although not very great, were yet not to be despised.

It need scarcely be noted that Mr. Cooper had long been a member of the Reform Club, ever since 1837 indeed, and some years before his death the appointment of Solicitor to the Club was conferred upon him. This was barely more than a graceful compliment to one who had always worked bravely for the Liberal cause, and as such he esteemed it.

Mr. Cooper's health began to fail him some three years before his death, when an attack of paralysis, from the effects of which he never entirely recovered, rendered him less capable of attending to his various professional and official engagements than before; but, save that his articulation had grown rather indistinct, his vigorous intellect survived with him almost to the last. At length, on the 28th of December, 1875, as before stated, he closed his eyes upon all that pertains to this world, to the deep grief of an only sister his affection towards whom, and towards his mother, who died in 1867 (his father having died in 1841 26 years earlier), was of the most devoted and self-sacrificing character. Two of his brothers predeceased him. His second brother, Dr. T. H. Cooper, lives to lament his loss, while numerous friends, between whom and himself a warm attachment subsisted, will long remember one whose place in their esteem cannot in all respects be easily filled up. Mr. Cooper was ever ready to lend, and very frequently did lend, a helping hand to any historical student or enquirer, who, modestly confessing his shortcomings, sought his assistance. But, woe betide the shallow boaster or empty pretender, who should attempt to display his accomplishments in his presence! Small mercy got he. The daw in borrowed feathers was soon denuded of his false plumage, and he submitted as best he could to the scarifying operation he had undergone. For the rest, like all men endowed with true humour, he was not only light-hearted but also large-hearted. Mr. Cooper was never married.

The interest taken by Mr. Cooper in the progress and success of the Sussex Archæological Society would, had he no further claims on its lasting remembrance, be sufficiently evidenced by the number and value of his contributions to its volumes, which, beginning in Vol. ii, were, with the exception of Vol. xi, continued through the whole series to Yol. xxv, inclusive. In Yol. ii. we have a paper on Papists and Recusants in Sussex in 1587, and another on Hastings Castle, Rape, and Town. In Vol. iii, an elaborate paper on The Lewknor Pedigree. In Vol. iv, Extracts from Account Books of the Everenden and Frewen Families. Queen Elizabeth's Visits to Sussex supply him with material for his paper in Vol. v, and to Vol. vi. he contributes a paper on the Liberties and Franchises within the Rape of Hastings. Vol. vii. contains his interesting paper On the Retention of British and Saxon Names in Sussex. Vol. viii. is enriched by his exhaustive paper on The Families of Braose (of Chesworth) and Hoo; and another entitled Notices of Winchelsea in and after the Fifteenth Century. To Vol. ix. his contributions are Annotations on Dr. Smart's extracts from the MSS. of Samuel Jeake; and Brambletye Chantry and Sedition in Sussex in 1579. In Vol. x. is his paper on Smuggling in Sussex, a paper curiously suggestive of the contrast between a state of things, the latter days of which some of our old South Coast dwellers yet living can remember, and the present. Vol. x. also contains a paper on Tokens Struck in Sussex in the Eighteenth Century. A paper on Proofs of Age of Sussex Families will be found in Vol. xii, which is supplemented by a short paper on the same subject in Vol. xv. The Oxenbridges of Brede Place, and of Boston, Massachusetts, one of his best articles, will also be found in Vol. xii. Vol. xiii. contains a List of Grants to Tipper and Dawe; another on Protestant Refugees in Sussex; and a third on Letters and Will of Andrew Borde. Vol. xiv. contains Notices of Hastings (a partnership paper by himself and Mr. Thomas Ross) and one On the Marriage Settlement of Isabella Poynings and William de Cricketot. In Vol. xv. have have the Poynings Pedigree; a paper on the Bonvilles of Halnaker; and a third of considerable interest on Sussex Men at Agincourt. In Vol. xvi. are papers on the Social Condition of the People in Sussex; and on Bramber, its Castle, Elections, andc. In Vol. xvii. he edits Mr. Sharpe's Notes on Ninfield and its Registers; contributes a paper by himself on and Supplies from Sussex; and is associated with Mr. Lower in a third paper, entitled Further Memorials of Seaford. In Vol. xviii. he has three papers, the first, one of considerable historical value, on the Participation of Sussex in Cade's Rising. The other two are respectively Notes on Sussex Castles, and Extracts from the Passage-book of the Port of Rye. This latter paper is followed up in Vol. xix, by one on Aliens in Rye, temp. Hen. VIII., and the same nineteenth volume also contains a paper on Royalist Compositions in Sussex during the Commonwealth. Vol. xx. is led off by a paper on Midhurst, its Lords and its Inhabitants. Vol. xxi. contains three papers, viz., Notes on Mayfield; Crown Presentations to Sussex Livings; and Additional Contributions towards the Parochial History of Hollington. Vol. xxii. has a paper on the Guilds and Chantries of Horsham;

and Vol. xxiii. Further Notices of Winchelsea, Former Inhabitants of
Chichester are chronicled in Vol. xxiv. Parham and its Collections forms
the commencing article in Vol. xxv; and this (save two inconsiderable notes
in Vol. xxvi.) is the final contribution from the indefatigable pen which
death alone could stay.

But the foregoing catalogue, full as it is, does not embrace all the printed
communications of William Durrant Cooper to the volumes of, nor does it
even refer to other valuable services rendered by him to, the Sussex
Archæological Society. Nearly a column, on pages 96 and 97, of the General
Index to our volumes, is devoted to his Minor Communications;
Information to other Contributors, andc., while during the years that he
officiated gratuitously as Editor of the Society's volumes, his multifarious
footnotes, as valuable as they are unobtrusive, attest at once to his industry,
his critical acumen, and the large extent of his historical acquirements. On
his retirement from the Editorship of our Society's volumes some of the
members (by a separate subscription), in order to mark their sense of Mr.
Cooper's services, resolved on asking his acceptance of some tangible
memorial of their gratitude and esteem. The result was a handsome silver
salver, engraved with a wreath of Sussex oak leaves and acorns, the Sussex
arms, and the family arms of Mr. Cooper." This memorial was presented to
Mr. Cooper at tho Society's Meeting, at Pulborough, in August, 1865, by the
hands of the late Bishop of Chichester, Dr. Gilbert, and it, of course, bore a
suitable inscription.

To the Camden Society's publications Mr. Cooper contributed as under: To
Vol. Iv. The Trelawny Papers, extending in date from 1644 to 1711, and
having reference to the famous West Country bishop, Jonathan Trelawny,
of Cornish celebrity. To Vol. lxxii. he contributed The Expenses of the
Judges of Assize riding the Western and Oxford Circuits, temp. Elizabeth,
1596-1601. From the MS. account-book of Judge Walmysley. Besides the
expenses of the Judges, these extracts contain "lists of the numerous
presents of provisions for their table, and of the places and persons which
entertained them." And Vol. lxxxii. which was the only entire volume of the
series edited by him, was, as Sussex readers know, on a subject he had
already made his own: Lists of Foreign Protestants, and Aliens, resident in
England, 1618-1688, from Returns in the State Paper Office. The
Introduction to this volume contains much valuable information.

In commenting on the loss the Camden Society had sustained in Mr.
Cooper's death, the Council speak of their departed colleague as a constant
attendant at their meetings, "always ready to contribute valuable advice and
criticism; his learning and his practical acquaintance with business will be
often missed by those with whom he so heartily co-operated in the interests
of the Society." Although a service-rendering member both of the Percy

Society, and the Shakespeare Society, Mr. Cooper would seem not to have been a contributor to the Percy Society volumes, while one volume only of the kindred Society claims him as its editor. The truth is that the staple commodity with which these Societies dealt belonged rather to the region of fancy than of fact, Still, the one volume for which the Shakespeare Society is indebted to him, Ralph Roister Doister, a Comedy, by Nicholas Udall. And the Tragedie of Gorboduc, by Thomas Norton and Thomas Sackville. With Introductory Memoirs. Edited by William Durrant Cooper, F.S.A. is one of the most valuable of the series, and the critical faculty is as well shown therein as in any of his historical pieces, while the Memoirs of Udall, Norton, and Sackville—this last a famous Sussex worthy—could hardly be improved upon.

To each of the four published volumes of the London and Middlesex Archæological Society, of which he was a Vice-President, Mr. Cooper contributed a paper: In Vol. i. he descants on The Parish Registers of Harrow-on-the-Hill, with special reference to the Families of Bellamy and Page. Notes on Uxbridge and its former Inhabitants are the subject of his paper in Vol. ii. and the Churches and Parishes of Saint James Garlick Hithe, and Saint Dionis Backchurch, both in the city of London, are the topics dwelt on in Vols. iii. and iv.

To the Kent Archæological Society's volumes he sent one paper only, which is printed in Vol. vii. of its series, but that paper, as a glance at its title will show, is an important one, John Cade's Followers in Kent: inasmuch as it dovetails in with his "Participation of Sussex in Cade's rising," in the eighteenth volume of the Sussex Collections. And it may be worth while to mention here that the late Mr. B. B. Orridge, with Mr. Cooper's assent, reprinted these Cade papers in his "Illustrations of Jack Cade's Rebellion, from Researches in the Guildhall Records, together with some newly found letters of Lord Bacon, andc., London, 1869."

To the Surrey Archæological Society, his one contribution is an Additional Note on a Deed relating to John Evelyn.

To The Reliquary, for April, 1862, he furnished an elaborate paper of considerable historical interest, entitled, Notices of Anthony Babington, of Dethick, and of the conspiracy of 1586.

Besides, and beyond, the above extensive bead-roll of Mr. Cooper's literary labours, there are, doubtless, several Papers and Essays, of which for lack of information and opportunity, no note has been, or can be, here taken. Enough, however, there is to show how continuously and conscientiously he worked. And the bulk of his several communications is in quite an inverse proportion to the painstaking research required for their production. No writer could possibly be more anxious than he was, even in his slightest contribution, to arrive at the absolute facts in any particular case. No second-hand authority satisfied him, if a primary one was to be got at, whatever the trouble it cost him.

Mr. Cooper's long connection with the Society of Antiquaries, and his communications to its Archæologia, are so felicitously treated by Frederic Ouvry, Esq., the learned and popular President of the Society, that, premising his mention of Mr. Cooper's election as a Fellow in March, 1841, it would be treason not to quote his actual language, as given in his Annual Address in April, 1876, slightly abridging it here and there.

"In adverting to the death of William Durrant Cooper, I speak of a friend of forty years' standing, of one whose many good qualities I warmly appreciated. He was one of the oldest, as he was assuredly one of the worthiest, members of our body. His first contribution to the Archæologia was laid before the Society in March, 1855. It is entitled Further Particulars of Thomas Norton, and of Stale Proceedings in Matters of Religion in the year 1581 and 1582. In May, 1856, he contributed Notices of the Plague in England, derived from the Correspondence of John Allix, in the year 1664-1669. In February, 1858, we find him reading a Memoir entitled Notices of the Tower of London, temp. Eliz. and the Horse Armoury, temp. Charles I. His most important contribution to the Archæologia closes the list. I refer to his Notices on the Great Seals of England, used after the Deposition of Charles the first, and before the Restoration, in 1660. The paucity of Mr. Durrant Cooper's communications to our pages must be attributed, not merely to the scanty leisure of an active professional life, but also to the large demands upon his time and pen, which were made by the Sussex Archaeological Society, to whose volumes his contributions are at once abundant and valuable. Of the services, however, which he rendered to this Society, his contributions to our Transactions would give a very inadequate idea. It is in the records of our committees that we shall find the proof of his zealous attachment to our body. Speaking as an ex Treasurer, I can bear testimony—which I am sure my successor in that office will endorse—to the thoroughness with which he executed his duties as a member of the Finance Committee, going carefully into every account submitted for examination, doing his utmost to promote the financial prosperity of the Society, a friend to economy as distinct from parsimony, and ever ready with criticisms and suggestions which I felt were always entitled to respect, as they came from a cool head and a warm heart."

Any addition to the above eloquent tribute would be superfluous. All that need be added is, that this Memoir would have been much less complete, but for the valuable aid rendered by Mr. Cooper's only surviving brother, Dr. T. H. Cooper; the Rev. Geo. Proctor, D.D.; John Smith, Esq. of the Lewes Savings Bank; Frederic Ouvry, Esq. Pres. S.A.; Thomas E. Gibb, Esq. Vestry Clerk, St. Pan eras, Middlesex; J. S. Smallfield, Esq. and his old Sussex friends, G. P. Bacon, Esq. Robert Crosskey, Esq. J.P. and John Clay

Lucas, Esq. F.S.A. to all of whom the heartiest thanks are here tendered. Mr. Cooper's portrait is unavoidably absent from these pages, for the too obvious reason, that none of a satisfactory character is in existence.

MARK ANTONY LOWER

Nat. 14 July 1813. On. 22 March 1876.

Every reader of Lockhart's Life of Sir Walter Scott, and of Charles Cuthbert Southey's Life of his famous father, Robert Southey, has lamented that the fragments of Autobiography which occupy the preliminary pages of those popular works break off at so early a period in the career of the two illustrious littérateurs whose lives are therein chronicled. Mark Antony Lower, a far humbler light in the literary firmament, also, to use a word which was a favourite with him, " endeavoured" a sketch of his career. The fragment which he has thus left behind him, is, alas! too brief to do more than exhibit its hero's advent upon life's threshold; but brief as it is, it is sufficiently interesting to induce a regret that its writer proceeded no further with it. Doubtless, both in the case of the eminent men above named, and in Mr. Lower's, the task of self-anatomization proved to be too painful to be persevered in. However, the outline of his life can hardly be initiated in a better way than by the presentation of his own story of its commencement:—

"RECOLLECTIONS OF A LITERARY LIFE."
By Mark Antony Lower.

"Eheu fugaces anni! How have the years fled since my life commenced, and my literary career began! At first sight it seems almost absurd for any man to sit down to the serious and laborious task of his own biography. As years increase, our years, our months, our weeks seem to become shorter. We seem to be as * of yesterday, and to know nothing.' Yet I never met with a man who, in spite of all his infirmities, his failures, his sins, would like to live his life over again: the probability being very strong that it would be merely a repetition of infirmity, of failure, and of sin. This is a wise arrangement of Heaven, for if the contrary feeling were indulged, and a redivivus were granted to men, ere long would the world become choke-full of Methuselahs, and forthcoming generations would have to migrate to the uninhabited planets, if, indeed, any such really exist.

"Still, the practice of writing men's lives, either autobiographically or by the pens of others, has prevailed from the very dawn of literature. The oldest written book extant informs us that 'there was a man in the land of Uz, whose name was Job,' and furnishes us with his history, and the opinions of himself and his contemporaries. Throughout the whole course of the Hebrew, the Classical, and the Middle Ages, down to our own days, a passion has existed for narrating the lives of men; and though the autobiographies are few in comparison with the 'memoirs' (as they are commonly called) yet by a critical examination of the works of poets and novelists we shall very often find, running through the thread of their writings, reminiscences of their lives, amounting almost to autobiographies. Of this we have eminent examples in King David's Psalms, in Horace, and in Oliver Goldsmith; perhaps, also, in Thackeray and Lord Lytton. Reminiscences will crop up in spite of ourselves, and we can no more prevent this phenomenon, than could the heroes of Trafalgar and Waterloo, of a few years since, be prevented from 'fighting their battles o'er again.'

"But what am I aiming at? Do I pretend to rank myself among the Davids, the Horaces, the Goldsmiths, the Lyttons, or with the hero of a hundred fights? Not so! My meaning is, as I advance to threescore years, to put upon simple record some of the events of a life which, though not altogether uneventful, has been that of a simple, unambitious man, a life which, though somewhat queer and picturesque, has not been marked by any deeds of a stirring or sensational character, though it may yet be worthy of record for the information of the coming generation. It contains incidents which may prove useful as lessons of what to do and what to avoid, and thus be regarded with some small amount of interest and profit.

"I was born in the obscure agricultural village of Chiddingly, in the Weald of Sussex, 14th July, 1813. My father, Richard Lower, was a schoolmaster of the old-fashioned middle-class of his profession. Without being what is called a scholar, he was a man of varied attainments. He had few associations except with farmers and tradesmen. There was in the somewhat extensive parish no resident squire or clergyman, and hence he became the factotum of the district. He was an excellent practical mathematician, and a land-surveyor of considerable note. He also held nearly every parochial office, made wills and agreements, and was an acknowledged authority in every local matter. Moreover, he was a capital self-taught draughtsman, and although his Latin was small, and Greek smaller still, he was one of the best English grammarians I ever knew.

Besides this, he was no mean poet, and every local event was by him chronicled in rhyme, and printed in local newspapers. In his eightieth year he published a small volume entitled ' Stray leaves from an Old Tree.' Speaking without prejudice, I can fairly say that few men in his sphere of life lived more usefully and more unselfishly than he did. Yet, with all his acquirements, which filled his rustic neighbours with astonishment— though

'..... still the wonder grew,
That one small head could carry all he knew,'—

he died at an advanced age as poor as when he commenced his useful existence.

"Under the guidance of this good father, I learned the rudiments of useful knowledge, and was soon an adept in most things that a young boy is capable of. Among the 'accomplishments,' I learnt music and drawing so early, that I cannot remember my first lessons in either science. I have not the slightest recollection of the hours when I learned my gamut, and a certain facility in sketching from nature. This I recollect, that I was a tolerable proficient on the flute, and a sketcher, before I was seven years old. The singing of sacred music was also one of our family amusements and recreations, and we frequently sang hymns set to music by my father himself. Those summer evenings that we spent in the garden, with our family, assisted by some musical neighbours and a few of the pupils, are a thing not easily 'disremembered.' A crowd of rustic neighbours behind the garden wall formed a well-pleased audience, and there we remained until the dews of nightfall warned us to retire to family prayers and to our peaceful couches each and all as tranquil and happy, and as unmindful of to-morrow's trials as ever the household of the Vicar of Wakefield could be."

It has been said, over and over again, that the life of a student and man of letters seldom affords much to relate of a personal character, and the life of Mark Antony Lower can hardly be said to be an exception to this rule.

Taking up his story at the point at which he may be presumed to have laid down his pen, and, noticing in passing, that he was the youngest of six sons, four of whom died in infancy, it may be mentioned that his first essay in the vocation which he made the business of his life during the greater part of it, was as an assistant to his sister, who opened a school at Easthothly, in 1830. He remained with her but a short time, for we find him attempting to establish a school for himself in the same year at Cade Street, in the parish of Heathfield, where he lodged during the week, returning home to his father's on the Friday or Saturday, as circumstances dictated. After spending some eighteen months on this experiment in the tutor's art, he removed, in his nineteenth year, to Alfriston, and there ventured on a more ambitious effort at school-keeping. And it was during his tenure of this Alfriston school that he enlarged his qualifications for teaching. In his scanty leisure at this time he made himself master of the Latin tongue, having, as he informed one of his oldest friends, his 'Latin grammar for sauce, while discussing his dinner.'

His hands must, indeed, at this period have been the reverse of idle, for it was then that he managed to bring before the public the first of the long series of literary works that bear his name, the title of which, in all its comprehensive fulness, is here given:

"Sussex: Being an Historical, Topographical, and General Description of every Rape, Hundred, River, Town, Borough, Parish, Village, Hamlet, Castle, Monastery, and Gentleman's Seat in that County. Alphabetically Arranged. With the Population of each Parish, according to the Census of 1821, and other useful and curious Information. With a correct Map of the County. By Mark Antony Lower. Printed for the Author, and sold by R. W. Lower, High Street, Lewes; W. Leppard, East Street, Brighton; and all Booksellers in the County. Mdcccxxxi."

Long years after this really well-compiled volume had been before the Sussex public, its author, grown fastidious by reason of his much larger acquaintance with topographic lore, has been occasionally heard to express his regret that he had ever published it. But it was, and even now is, although thrown into the shade by his larger and much more recent work on the same subject, still a very serviceable compilation, and one that the writer, however popular in his later days, need not have blushed at being identified with. And that he must even then, by some proofs given of literary aptitude, have acquired, comparatively speaking, considerable local repute, is evidenced by the patent fact that the subscription list appended to his book, comprises the names of upwards of 250 patrons.

Before he finally gave up his school at Alfriston, he set about to establish a Mechanics' Institution there, a meritorious and successful work, in which he was largely aided by John Dudeney, a name never mentioned in East Sussex but in terms of esteem and admiration. It would be foreign to the purpose of this memoir to dwell here upon the character and career of this "hereditary Southdown shepherd;" but how, springing from humble parentage, taught merely to read by his careful mother, he ultimately lifted himself up from his pastoral occupation to become a successful schoolmaster, and the founder of a Philosophical Society in Lewes, a practical and accomplished naturalist, and an instructive and amusing lecturer on Astronomy, is told, not only in John Dudeney's simple English, in his own account of himself, in the Second volume of the Sussex Archæological Collections, but also by his attached friend, Mark Antony

Lower, in the Gentleman's Magazine, for March, 1853. It is hardly needful
to add that the friendship between these two, which began at Alfriston, was
severed by death only, and that they were closely associated in continuous
efforts towards the elevation and education of the working classes.

Somewhere about half way between Alfriston and Chiddingly there stood,
and probably still stands, a pleasantly situated farmhouse, with the
comfortably circumstanced occupants of which, in his weekly or more
frequent walks to and from his home—for he had not cut himself entirely
away from the parental roof—our handsome young dominie became
acquainted; and there, on scorching July, or freezing January afternoons, a
welcome rest of half-an-hour or more.was often brought to a close by an
acceptable cup of tea, with its appropriate accompaniments. Nor was the
latter the sole or chief charm of his haltings at the domicile of this estimable
family. A bonny, bright-eyed, flaxen-haired young maiden, who officiated as
governess there, soon brought home to Mark Antony Lower the conviction
that he possessed a susceptible heart. She who thus enmeshed that heart of
his in golden and enduring fetters, was of a well-known and still flourishing
Sussex family, the Holmans, and it may be well to anticipate chronological
events, by noting here that Mark Antony Lower and Mercy Holman became
husband and wife at Bromley, in Kent, in the year 1838.

In or about the year 1835, Mr. Lower, by that time confirmed in his liking
for his chosen vocation, and more fitted for it by his four or five years'
experience in his village school ventures, removed to Lewes. He there
"hired an old chapel or rather preaching room, close by the then
Lancastrian School, but now the British School, in Lancaster Street." He
soon gathered round him a goodly number of scholars, and made such
satisfactory progress, that in due time he felt warranted in giving up his
bachelor lodgings, and going into housekeeping, taking unto wife, as already
mentioned, the above-named Mercy Holman: and fortunate was he, and
fortunate he ever deemed himself, in drawing so unquestionable a prize in
the matrimonial lottery.

Previously, however, to making this important jump in life, and full of
enthusiasm, as yet uncooled, he took a prominent part in the establishment
of "the Lewes New Temperance Society," the first annual Report of which,
a document which need not be quoted here, proceeded from his pen.
Subsequently, for reasons cogent enough to him at the time, he saw fit to
secede from this body, but he did not relax in his other efforts for the
mental advancement of the masses.

His successive removals from house to house, as both his school and family
increased, require not to be chronicled in detail. But in connection with one
of these dwellings an anecdotical incident may be related. In the garden of a
next-door neighbour grew a handsome pear tree, which, in proper season,

bore a full crop of fruit, and stood in tempting proximity to, in fact overhung, the dividing wall, between the school play-ground and the aforesaid garden. "Where is the school-boy who would not covet his neighbour's goods when, day after day, they thus as it were stimulated the desire of possession and enjoyment? Stone throwing was resorted to, a fall of fruit resulted, but, unfortunately, broken windows resulted also. The neighbour complained of his fractured glass; the schoolmaster apologised, promised punishment of the offenders, and offered to send in a glazier to repair the damage done. "No!" said the good-natured sufferer, " I will take that charge upon myself, and, by way of truce with your young scapegraces, I will send in a basket of pears annually, and they shall covenant to do my trees and windows no further harm!" The point of the story is, that when the pears were sent in, as agreed on, the ringleaders in the mischief claimed the greater share, on the ground that but for their pluck at the outset, the gift of fruit would never have been made!

Most dwellers in, and many sojourners at, Lewes, have witnessed, or heard of, the mad pranks of the self-styled "Bonfire Boys," who, on the Fifth of November every year, startle the quiet old town from its accustomed propriety, and, by their uproarious proceedings, put nervous wayfarers in fancied danger of their lives. Mr. Lower was much opposed to this "Saturnalia of the roughs" and, on one occasion, under the signature of " A Young Inhabitant," he issued a printed manifesto, very earnestly worded, deprecating the continuance of the irrational and mischievous custom; especially warning its perpetrators against the consequences of their reckless flinging about of squibs and other fiery missiles, and noting that, "in cases of fire happening on such occasions, the insurance companies are not responsible for the loss." The authorship of this broadside was soon bruited about, and the mob threatened to throw its writer into the river; but he prudently kept out of the mob's way, and to this day the Lewes Bonfire Boys, now recruited by reinforcements from the riff-raff of Brighton, make "night hideous" once a year on poor Guy Fawkes's Anniversary, to the terror of all peaceable folk within hail of their doings.

About 1853 or 1854 Mr. Lower removed to Saint Anne's House, his last and longest inhabited dwelling in Lewes, an old red-brick edifice, of somewhat irregular character, formerly occupied by, among other local celebrities, some of the Shelleys, by Sir Roger Newdigate, founder of the Newdigate prize at Oxford, and other locally distinguished persons. Still earlier it was the property and home of John Rowe, whose name is held in reverence as that of the Father of Sussex Archæology. Such a domicile became, therefore, the appropriate abiding-place of so eminent a student of past times as Mark Antony Lower. The house has now disappeared, and on its site a modern, and doubtless much more convenient, residence rears its head, but the old Saint Anne's house was associated with times and men round which and whom the halo of Antiquity has long gathered. Mr. Lower occupied this ancient house until the year 1867. In it were written the greater part of his many papers and books, and within its walls he continued to pursue his scholastic vocation, limiting his pupils to boarders only, among whom were, generally, several young Frenchmen, for whose tuition he specially laid himself out. But, towards the latter years of his stay in Lewes, the establishment of public and semi-public schools and colleges for the sons of middle-class parents, as also the failing health of himself and his devoted wife, upon whom devolved the domestic superintendence of his modest academy, told upon the number of his inmates, and his consequent pecuniary returns, when, in the last-mentioned year, Mrs. Lower succumbed to the malady under which she had been suffering, and her husband, in the thirtieth year of his wedded life, a life which, in regard to his domestic happiness, had been all that he could have desired or anticipated, found himself a bereaved widower, at a season when he could ill spare so loved and loving a partner. To a man of his strong affections, this melancholy event was productive of considerable distress of mind, and a few months afterwards, under the altered circumstances in which he found himself placed, he sold his dear old house, broke up his school, and removed to Seaford, still taking a few French pupils.

Anxious to mark their sense, and appreciating the importance of Mr. Lower's long labours in connection with the history of his native county, his friends—members and non-members alike of the Sussex Archaeological Society—organized a committee, and raised a subscription, in testimony of the high esteem in which they held him and his services. The fund thus raised, amounting to about 400, was presented to him some short time before his removal to Seaford.

Some three years after this removal he took for his second wife a maiden lady whom he had long known—Miss Sarah Scrase—of an old and respectable family, originally Danish, long settled in Sussex. Soon after this second marriage, namely, in 1871, he left his native county entirely, to reside in London or its immediate neighbourhood, in order to apply himself, as far as his impaired health would permit, to literary pursuits.

In 1873, a trip to Denmark and Sweden was resolved on, partly under the hope that his health would be benefited by it, and partly with the object of pursuing some inquiries, of an archaeological character, among a people so nearly allied to our own in several important particulars. His wife accompanied him. But the health-seeking pilgrimage failed of its object, and he was, after too short a sojourn for his literary purposes, ordered back to England by his physician, by the most expeditious route. A book, however, the last his hitherto active hand produced, was the outcome of his otherwise fruitless journey, and this book, Wayside Notes in Scandinavia.

London, 1874, presented, it must be confessed, but a faint reflex of his usual lively style of composition.

In 1875 it was his misfortune to follow his second wife, who was affectionately attached to him, to the grave. After her death he removed from his abode in the southern suburb of London, Peckham, to the house of his youngest daughter, Mrs. Hawkins, at Enfield, Middlesex, where, surrounded by such of his six surviving sons and daughters as happened to be in England, he passed away, on the 22nd of March, 1876, in the sixty-third year of his age.

Mr. Lower took no active part in the municipal affairs of Lewes; he served as one of the Headboroughs in the year 1860-1861, but never held any other prominently public office.

It is as one of the originators of the Sussex Archæological Society, and one of the chief co-operators in its valuable Collections both by his pen and pencil, that the name of Mark Antony Lower will ever deserve to be remembered with the highest honour. And as he was one of the pioneers of the movement, so was he the last survivor of the six coadjutors, Messrs. Blaauw, Blencowe, Dudeney, Figg, Harvey, and himself, who, by their deliberations, gave currency to an idea first started, it is believed, and gradually worked out, in the frequent neighbourly meetings at each other's house alternately, of the last-named four. This view is borne out by the Report prefixed to the first volume of the Society's Collections, where it appears that " the first meeting which defined the objects and established the rules of the Society, took place on June 18th, 1846, at the suggestion of a few gentlemen in the town and neighbourhood of Lewes, who, observing the interest excited by some recent antiquarian discoveries, were anxious to promote a readier acquaintance among persons attached to the same pursuits, and to combine their exertions in illustration of the History and Antiquities of Sussex." So, with the Duke of Richmond, Lord Lieutenant of the County, as patron, and the Duke of Norfolk as President, the Sussex Archæological Society speedily became an "accomplished fact."

The first public meeting of the Society was held on July 9th, 1846, in an appropriate arena, the ruins of Pevensey Castle; and the first paper read there, was on the History of those venerable remains, by Mark Antony Lower, which paper, under the title of Chronicles of Pevensey, was published by its Author as a separate work, and has since passed through several editions as a popular handbook for visitors.

Noting the fact, that the first volume of the Society's Collections opens with a paper on the germane subject of Sussex Archæology, from the scholarly pen of the late Mr. Blaauw, who then officiated as honorary secretary, the several contributions of Mr. Lower to those collections now claim especial notice.

In Vol. i. we have three papers, the subjects of which respectively are:—

Seals of the Sussex Cinque Ports; Names of the Sussex Gentry in 1588, a short, but very suggestive paper, as supplying the names of upwards of 100 Sussex contributors to the 'extraordinary aid' called for by Queen Elizabeth on the threatened invasion of England by Spain; and An Ancient Leaden Coffer found at Willingdon. Vol. ii. in addition to five shorter papers, namely, Observations on the Landing of William the Conqueror; On Oliver Cromwell's Pocket Bible; On a Congratulatory Letter to Sir Thomas Pelham; On Roman Remains at Eastbourne, and, On the Monumental Brasses of Sussex, contains his long, valuable, and remarkably able paper On the Iron Works of Sussex, copiously illustrated by drawings from his own hand. The universal interest excited in the Iron districts all over England by this paper, led to the speedy sale of all the copies of the volume containing it, and the second-hand booksellers print "very rare" against it in their catalogues, when lucky enough to get hold of a copy. Mr. Lower contributed two supplementary articles to Vols. iii. and xviii. on the same subject, Vol. iii. exhibits three articles: On the Castle of Bellencombre, in Normandy; On Wills proved at Lewes and Chichester; and On the Pelham Buckle and De la Warr Badge. In Vol. iv. we have his amusing papers On Sir Bevis of Hampton, and his Horse Arundel; On some Wills of Inhabitants of Herstmonceux and neighbouring parishes; and a third, prettily illustrated by his own careful drawings, On the Star Inn at Alfriston. Vol. v. opens with a paper by him On the Descent of Wiston, with Anecdotes of its Possessors; while a second is On Miscellaneous Antiquities discovered in and relating to Sussex; and a third On Watermills and Windmills in Sussex. His first paper in Vol. vi. is on the stirring theme of the Battle of Hastings. A second is entitled Memoranda of the Boord or Borde Family, with a Memoir of Andrew Borde—the original of all the Merry Andrews of our old country fairs—and the third is on a subject he had already made his own Pevensey Castle and recent Excavations there. Vol. vii. contains his long and painstaking Memorials of the Town, Parish, and Cinque-port of Seaford, an account which is supplemented in Vol. xviii. by a joint paper by himself and Mr. Cooper, entitled Further Memorials of Seaford. A Genealogical Memoir of the Family of Scrase is his only contribution to Vol. viii. Five articles from his fertile pen are to be found in Vol. ix. viz. Notes of the Family of Miller, of Burghill, and Winkinghurst; On the Churches of Newhaven and Denton; Notes respecting Halnaker, Boxgrove, andc. temp. Q. Eliz.; On the Pillory and Cucking Stool at Rye. But the gem of these five papers is that entitled Bodiam and its Lords, a charming contribution, and one which, some half-dozen years ago, he revised and republished, at the instance of the present owner of that dismantled stronghold of the Dalyngruges, still, in its ruins, a picturesque and majestic pile. A chatty paper On certain Inns and Inn Signs in Sussex, appears in Vol. x. and Extracts from the Diary of a Sussex Tradesman 100 years ago, edited by himself and Mr. Blencowe, is all that bears his name in Vol. xi. In Vol. xii. his paper On the Hospital of Lepers at Seaford, is followed by Notices of Sir Edward Dalyngruge the Builder of Bodiam Castle, a pendant to the Bodiam paper above referred to. In Vol. xiii. he gives us the Will of a Sussex Clergyman 300 years ago, and a paper on a subject he had made himself peculiarly master of, Old Speech and Manners in Sussex. To Vol. xiv. he contributed his Parochial History of Chiddingly, his native parish, be it remembered, and a model of the familiar style in which such a subject should be treated. In Vols. xv. and xvi. we have an exhaustive account of The Rivers of Sussex. The author's masterly handling of his aqueous topic earned, in this instance, the praise of that most critical of critical journals, the Saturday Review. Notes on Jack Cade and his adherents, an acceptable addition to Mr. Cooper's paper, is the first of four contributions to vol. xviii, the other three being, a Catalogue of Antiquities in the Society's Museum, Lewes Castle (jointly with Mr. E. Chapman), Notes on Sussex Castles (in which Mr. Cooper was his coadjutor), and On a "Kitchen Midden" at Newhaven. His quota of papers in Vol. xix. also numbers four, namely, On some old Parochial documents relating to Lindfield; Notes on the Family of Whitfeld, or Whitfield, of Northumberland and Sussex; an account of the tragic poaching affray which ended in the Trial and Execution of Thomas Lord Dacre, of Herstmonceux Castle, for Murder; and a brief essay On the Tomb of Richard Burré in Sompting Church. His single contribution to Vol. xx. is On Sir William Springett and the Springett family. He and the Rev. Edw. Turner together furnish Parochial Notices of Horsted Parva to Vol. xxi. In Yol. xxii. his pen is employed On Deeds of the Ancient Family of Cobbe and others, of Sussex, relating to property in Arlington. Notes on old Sussex Families supply him with an apt theme for two papers in Vols. xxiv. and xxv. In Yol. xxiv. he has two other papers; one entitled Newspaper Cuttings relating to Sussex (1678-1771), with Notes and Observations; the other On the Norman origin of the Family of Pelham. In Vol. xxv. appear Some Notices of Charles Sergison (temp. William III. and Queen Anne) and (jointly with Mr. Elwes) Additional Notices of South Bersted. In Vol. xxvi. a Translation of a Latin Roll relating to the Liberties and Immunities of Battel Abbey, the joint work of himself and Mr. J. R. Daniel-Tyssen; and a short paper On a Miniature of John Selden, bring to an end the tale of his chief contributions to the Sussex Archaeological Collections.

For the smaller matters grouped under the head of Minor Communications, Information furnished to other Contributors, andc. seeing that Mr. Lower stands credited with an aggregate of items filling nearly two columns in the General Index, they cannot be enumerated here. Indeed, " their name is Legion." Nor must it be forgotten that as, in his own words, he " was a

sketcher before he was seven years old," he was enabled to render good
service in a double capacity, as his numerous, and at once faithful and
effective drawings, in several of our volumes bear picturesque witness.
Moreover, he not only supplied the pictorial illustrations to most of his own
articles, but he likewise illustrated the articles of some of his fellow-
contributors. Practically, he seldom had the heart to say "Nay!" to any call
made upon him, in the direction of his favourite pursuit, whatever sacrifice
of time or labour it might entail; and, too often, he allowed strangers to
seduce him from his school interests, when a strict regard for them ought
to have forced from his lips the utterance of the negative monosyllable.

On Mr. Cooper's retirement from the Editorship of the Society's
Collections, in 1865, the Committee, "considering the propriety of
appointing a Salaried Editor and Corresponding Secretary," resolved that
"Mark Antony Lower, Esq. F.S.A., be appointed" to the joint office, "with
such remuneration as the Committee may think his time and labour
demand." In 1870 his continued ill-health compelled him to retire from this
office, on which occasion the following resolution was entered on the
Minutes of the Society's proceedings:—

"Having accepted the resignation by Mr. Lower of his office of Editor of
the Society's Collections, the General Committee desire to place on record
their appreciation of the services rendered by him to the Society. In the
establishment of the Society Mr. Lower took a prominent and very useful
part; in the general conduct of its affairs he was ever most zealous; and
every volume of the Collections hitherto published, contains evidence of his
wide knowledge and research, in his character both of Contributor and
Editor. The Committee have great pleasure in knowing that though Mr.
Lower has resigned the office of Editor, his valuable co-operation will not
be wholly withdrawn."

Of his principal separate publications, the title of the earliest has been
already set out in full, on a preceding page, as also that of his latest. For the
remainder the following list may be taken as tolerably complete:—

English SURNAMES. Essays on Family Nomenclature, Historical,
Etymological, and Humorous. "With Chapters of Rebuses and Canting
Arms, the Roll of Battel Abbey, a List of Latinized Surnames, andc., andc.
By Mark Antony Lower. "What's in a name?" London. John Russell Smith.
MDCCCXLII. 8vo. A second edition being soon called for, the author
issued one, revised and enlarged. This was followed by a third edition, still
further enlarged, in two volumes, in 1 849; and, not long before his death, a
fourth edition, again with additions by the author, was published by Mr.
John Russell Smith.

The Curiosities of Heraldry. With Illustrations from Old English Writers.
By Mark Antony Lower.. With numerous Wood Engravings. From Designs

by the Author. London: John Russell Smith. MDCCCXLV. 8vo.

The Chronicle of Battel Abbey, from 1066 to 1176. Now first translated, with Notes, and an Abstract of the Subsequent History of the Establishment. By Mark Antony Lower, M.A. London: John Russell Smith. MDCCCLI. 8vo.

Contributions to Literature, Historical, Antiquarian, and Metrical. By Mark Antony Lower, M.A. F.S.A. London: John Russell Smith. MDCCCLIV. 8vo.

PATRONYMICA Britannica, a Dictionary of the FAMILY NAMES of the United Kingdom; Endeavoured by Mark Antony Lower, M.A. F.S.A. London: John Russell Smith. MDCCCLX. This work has a portrait of the author, somewhat too leonine, perhaps, but still very like; and a gracefully engraved border on the title page, from his own design. The characteristic portrait, here referred to, thanks to the courtesy of Mr. John Russell Smith, who kindly lends the wood block for the purpose, forms the frontispiece to the present memoir.

The Worthies of Sussex: Biographical Sketches of the most Eminent Natives or Inhabitants of the County, from the earliest Period to the Present Time; with incidental Notices, illustrative of Sussex History. By Mark Antony Lower, M.A. F.S.A. Printed for subscribers only. Lewes: G. P. Bacon. MDCCCLXV. Large 4to.

A Compendious History of Sussex, Topographical, Archaeological, and Anecdotical. Containing an Index to the first Twenty Volumes of the Sussex Archaeological Collections. By Mark Antony Lower, M.A. Lewes: G. P. Bacon. 1870. Two volumes. 8vo.

Historical and Genealogical Notices of the Pelham Family. By Mark Antony LOWER, M.A., F.S.A. Privately printed. 1873. Folio. Of this handsome example of typography from Mr. Bacon's press, a very few copies only were printed.

Sundry smaller, but not unimportant, publications merit a short notice, such as his Handbook for Lewes, which, first issued in 1846, has since passed through several editions. Then, for Prince Louis Lucien Bonaparte's renderings of the Song of Solomon into the various provincial dialects, he furnished a version in the Sussex vernacular, a task for which he was well qualified, and in which he succeeded to the full satisfaction of the Prince.

His Stranger at Rouen, a Guide for Englishmen (it can be bought in London, of Mr. Russell Smith) is a little book well adapted to its unambitious purpose. The descriptive text to Nibbs's Churches of Sussex is also from his pen.

Another little book bearing his name, and entitled The Sussex Martyrs, their Examinations and Cruel Burnings in the time of Queen Mary, comprising the interesting personal narrative of Richard Woodman, andc. andc. is a reprint of old John Fox's account, with a preface, and some elucidatory notes.

For his old friend, the London publisher of all his important works, Mr.
John Russell Smith, he edited The Lives of the Duke and Duchess of
Newcastle, by Margaret Duchess of Newcastle, and Camden's Remains
concerning Britain. And he contributed several articles to the same
publisher's Retrospective Review, a meritorious periodical deserving a much
larger share of patronage than, during its too brief existence, the wayward
English public chose to bestow upon it.

A work on the Bayeux Tapestry remains in manuscript.

Peculiar to Mark Antony Lower, was his thorough humanity, and his sense
of the humorous. Whatever the theme he enlarged upon, it went hard with
him if he could not find a human, or a humorous, side to it. His delight was
to gather up, in the highways and byways, the nooks and corners, of his
native county, quaint bits of character, anecdotes, and incidents of old
times, such as were calculated to throw light upon the social history of past
days. His humour, too, was a part of him, not an acquired faculty. The son
of the author of "Tom Cladpole's Jurney to Lunnun," and "Jan Cladpole's
Trip to 'Merricur, in search arter Dollar trees," he unquestionably inherited
from his sire his appreciation of the oddities and eccentricities of life in
every phase in which they were to be found.

But he was too honest and too earnest a student of antiquity to subordinate
reality to romance. Like his old friend Cooper, he was not over-enthusiastic
upon the subject of Prehistoric Archæology. The "Flint flake " and
"Kitchen-midden" theories found little favour in his eyes, and, in his paper
on the discovery at Newhaven of a so-called Kitchen-midden, in the
eighteenth volume of the Society's Collections, his incredulity relative to the
deductions of the Anthropological experts on that occasion, is, perhaps, a
little too pronouncedly expressed, and he recounts a dinner-table joke, got
up at their expense, with evident delight. He had not the same facility as his
confrere, above-named, who was domiciled in London, had, for consulting
authorities, of every kind, at the fountain head, and, sometimes, his forced
reliance on second-hand sources of information may have misled him, but
he shared his friend's anxiety to be correct. He lived in, and he loved, the
country; and so "racy of the soil" was he, that it was difficult to induce him
to sleep more than a single night in London, except under pressing and
unusual circumstances.

From a graphic article, entitled Through Sussex, in the Temple Bar
Magazine for January, 1866, the following passage will well bear
transplanting to these pages:—

"Lewes has a famous Antiquary the great authority on surnames Mr. Mark
Antony Lower. He is a gentleman with more poetry in him than most of
the Dryasdust School: witness his picturesque presentment of the Sussex

villages—clusters of lowly habitations, some thatched, some tiled, some abutting the street, some standing angularly towards it, all built of flint or boulders. A barn, a stable, a circular pigeon-house, centuries old, with all its denizens (direct descendants of the old manorial pigeons which lived here in the days of the Plantagenets), and an antique gable or two, peer out among the tall elms.' We fancied we met Mr. Lower close by Lewes Castle. I sketched on the margin of my Murray the ample forehead of the unknown, beneath an archaic hat, the keen observant eyes behind archaic spectacles; and shall leave it by will to the Sussex Archæological Society."

When in his prime, his constant devotion to his work, scholastic, literary, archaeological, kept him too much, it may be, engaged and, always talking about the holidays he meant to, but did not, take, when his school vacations arrived, one who knew his habits, Mr. Joseph Ellis, of Brighton, who, as his special intimates only know, is an admirable inditer of good-humoured flings at the amiable foibles of his acquaintances, "poked his fun" at his Lewes friend after the following facetious fashion:—

Mark Antony Lower enjoys his vacation,
But says there's no time in it for recreation !
And then, for long months, he pursues his vocation,
Like horse in a mill, without any cessation ;
Hence a problem involving no small botheration,
Namely:—which is Vocation, and which is Vacation ?
For the difference here between vo and va,
Should value the same as between work and play,
Or even as much as between do and say.
But whether in vo, or whether in va,
Or whether in work or whether in play,
Or whether in do, or whether in say,
The metamorphosis is with O and A:—
So with Lower—a slave who ne'er kicks off his fetters—
Call it work, call it play, 'tis a question of Letters!

A most obliging disposition; a sensitiveness well nigh feminine in its nature; a keen perception of the ludicrous; a ready hand at turning a pun or an epigram; and a happy way of rendering the anecdotes, wherewith his memory was copiously stored, made Mark Antony Lower always a welcome companion in the social circle. But less bright days came upon him. His closing years were darkened by impaired health, the sun of fortune shone but fitfully upon him, and continuous literary labour became at length an impossibility. The once robust figure had fallen away to such an extent, that some who knew him intimately, but who had not seen him for an interval of twelve months or more, failed to recognise at once their old friend in the

wasted form before them. The date of his passing away has already been
given. It may truly be added, that the void left in the ranks of Archæology
by his death, cannot, in the many-sided gifts with which he was endowed,
be easily filled up.

Mr. Lower was for several years a Fellow of the Society of Antiquaries of
London. He was Master of Arts of one of the United States Universities; a "
Fellow of the Societies of Antiquaries of Normandy, America, Newcastle-
upon-Tyne; and a Member of the Academy of Sciences of Caen."

Thanks are especially due to Mr. John Dudeney, of Milton House, Lewes,
as also to Mr. W. de Warenne Lower, second son of Mr. Lower, for their
important assistance in furnishing materials for the foregoing memoir; as
also to Mr. Lower's old pupil, Mr. J. E. Price, F.S.A.; to Mr. Joseph Ellis,
and to Mr. John Russell Smith.

www.ingramcontent.com/pod-product-compliance
Lightning Source LLC
Chambersburg PA
CBHW070755180526
45168CB00004B/1616